S0-ANR-671

JELLYFISH

UNUSUAL ANIMALS

Lynn M. Stone

The Rourke Corporation, Inc.
Vero Beach, Florida 32964

Edited by Sandra A. Robinson

PHOTO CREDITS
© Herb Segars: cover, pages 7, 13, 15, 17, 18; © Breck Kent: title page; © Frank Balthis: page 4; © Mary Cote: pages 8, 12; © Lynn M. Stone: pages 10, 21

Library of Congress Cataloging-in-Publication Data

Stone, Lynn M.
Jellyfish / by Lynn M. Stone.
p. cm. — (Unusual animals)
Includes index.
Summary: Discusses the physical characteristics, habits, and dangerous aspects of jellyfish and describes different species and their relatives.
ISBN 0-86593-284-0
1. Jellyfishes—Juvenile literature. [1. Jellyfishes.] I. Title.
II. Series: Stone, Lynn M. Unusual animals.
QL377.S4S76 1993
593.7—dc20 93-19462
CIP
AC

TABLE OF CONTENTS

THE UNUSUAL JELLYFISH

A jellyfish is an unusual creature indeed. You can almost see through a jellyfish—it is nearly **transparent,** or clear.

The jellyfish is not a fish at all. It belongs to a group of soft, boneless sea animals. A jellyfish is basically a floating blob with a mouth surrounded by a "beard" of fleshy **tentacles.**

The tentacles are unusual, too. They look harmless, but they pack a deadly sting.

Sea nettles are jellyfish with long, stinging tentacles

HOW JELLYFISH LOOK

Many jellyfish are shaped like bells or umbrella tops. Their jellylike bodies may be nearly clear, or a color such as pale blue, orange, brown or white.

Jellyfish may be tiny or quite large. Many of the most common kinds, or **species,** are saucer size. One species in the cold Arctic seas is huge. Its body can be more than 7 feet wide, and its tentacles can be 120 feet long—longer than a basketball court!

Moon jellyfish are common bell-shaped jellies

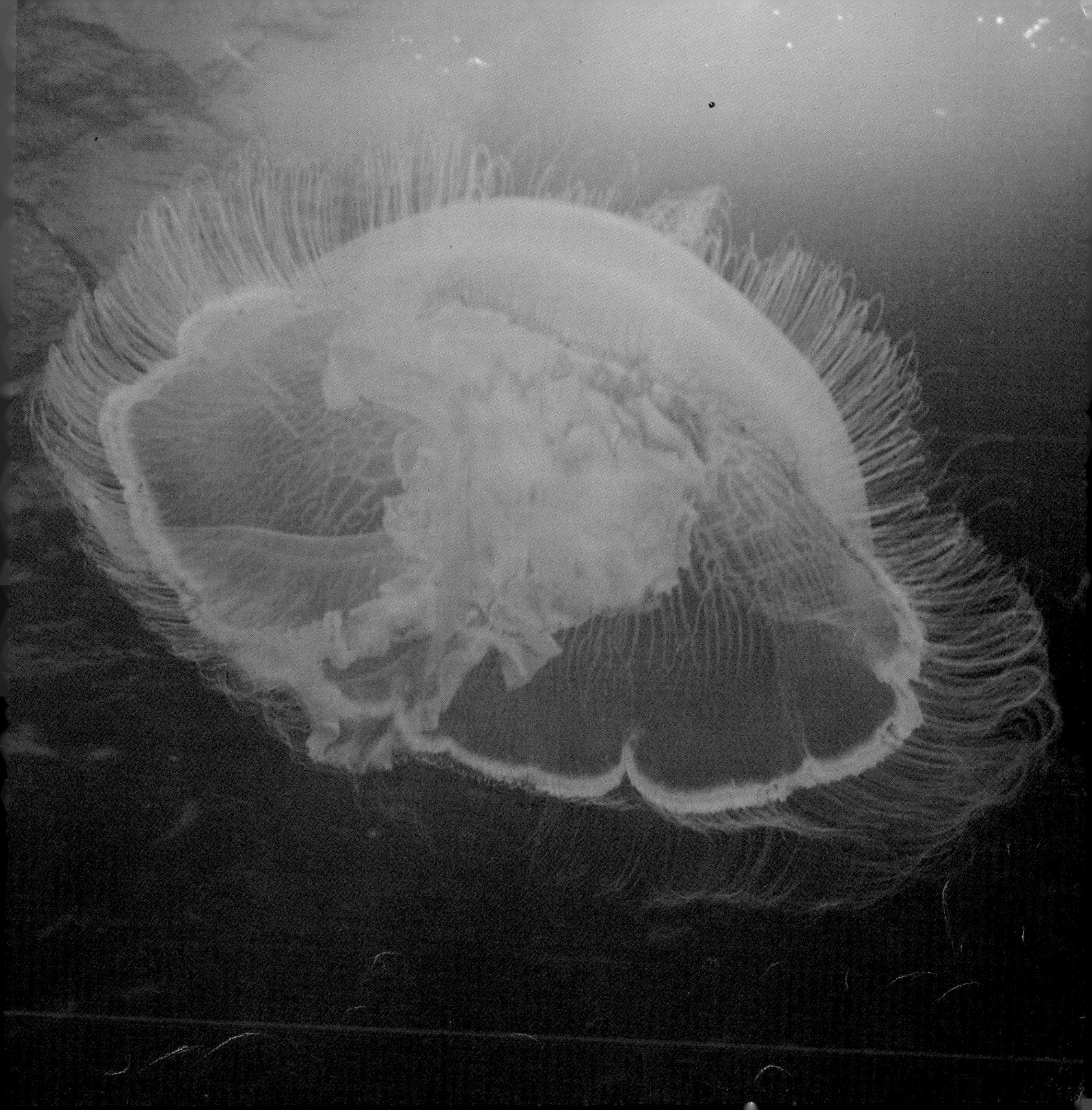

KINDS OF JELLYFISH

Hundreds of species of jellyfish live in the oceans of the world. The common moon jellyfish has short tentacles and a four-leaf clover pattern in its body, or bell.

The cannonball jellyfish is also known as the jellyball.

The upside-down jellyfish has a lacy fringe of tentacles. On sea bottoms it sometimes lies upside-down to soak up sunlight.

An upside-down jellyfish

JELLYFISH COUSINS

Jellyfish are related to two other well-known sea, or **marine,** animals—the **sea anemones** and corals.

Like jellyfish, anemones and corals are soft, simple animals that have bodies with just one opening—their mouth. Corals and anemones also have a fringe of tentacles around their mouths.

While jellyfish are animals that usually drift with the sea, corals and anemones live in one place.

Sea anemones, like jellyfish, are soft, simple animals with tentacles around their mouths

Another jellyfish relative, orange coral, feeding at night in the Caribbean Sea

Largest of the jellyfish, a lion's mane floats near the surface of the North Atlantic Ocean

PREDATORS AND PREY

Jellyfish dine mostly on little fish and small, boneless animals. Some of the animals jellyfish eat are **microscopic**—too small to be seen without a microscope.

Jellyfish themselves are **prey,** or food, for several kinds of animals. Certain fish and snails nibble on jellyfish tentacles. Some kinds of sea turtles also eat jellyfish. The animals that catch and eat jellyfish are called **predators.**

Cunners nibble a young lion's mane jellyfish

JELLYFISH TENTACLES

The stinging tentacles of jellyfish often trail from the animal like strands of hair—or snakes. Scientists call jellyfish medusa. In old Greek stories, a woman named Medusa wore a hairdo of snakes.

Tentacles are used to catch prey. The sting in tentacles comes from microscopic "darts" that are usually "shot" when the tentacle brushes against an animal. A jellyfish can shorten its tentacles to lift prey to its mouth.

Unharmed, young butterflyfish rest safely in a jungle of jellyfish tentacles

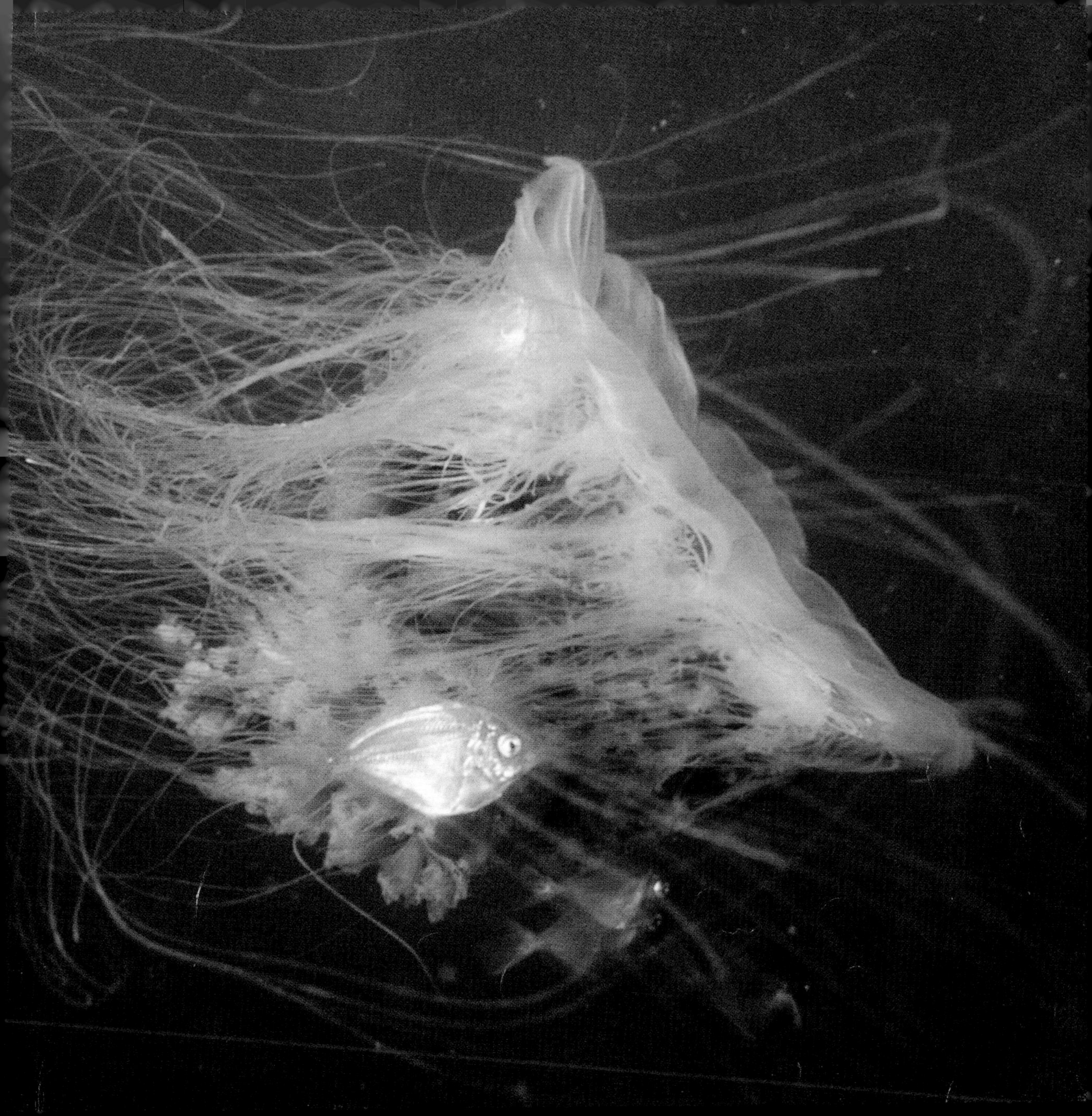

JELLYFISH HABITS

Jellyfish grow from eggs. As an adult, a jellyfish moves slowly through the sea by pumping water in and out of its body. A swimming jellyfish has little control over where it goes when it is swimming. It travels in the general direction that ocean currents take it. Strong winds blow thousands of jellyfish onto shores.

Some jellyfish, such as the sea wasp, can travel in deep water. Others float near the surface.

Winds and ocean currents often blow jellyfish, like this many-ribbed jelly, ashore

THE PORTUGUESE MAN-OF-WAR

The Portuguese man-of-war and its cousins are often mistaken for jellyfish. They are not true jellyfish, but they have tentacles and are closely related to jellyfish.

Instead of having jelly bodies, these creatures have air-filled bodies and float on the sea like balloons. The Portuguese man-of-war has a long, delicate beard of tentacles that may trail 30 feet behind the "balloon."

Another jelly cousin, the by-the-wind sailor, has a soft, triangle-shaped sail that catches sea breezes.

On Florida's Atlantic shore, the violet balloon float signals this is a Portuguese man-of-war

JELLYFISH AND PEOPLE

Swimmers should always avoid jellyfish. The tentacles of many jellyfish species can cause painful and sometimes serious injuries.

Jellyfish and their cousins cause more human injuries than all other marine animals together! The deadly tentacles of sea wasp jellyfish killed 30 people during a 25-year period in northern Australia. Sea wasps have a poison more deadly than snakes.

Glossary

marine (muh REEN) — of or relating to the sea, salt water

microscopic (my kro SKAH pik) — able to be seen only through the powerful lens of a microscope; invisible without a microscope

predator (PRED uh tor) — an animal that kills other animals for food

prey (PRAY) — an animal that is hunted for food by another animal

sea anemone (SEE an M un ee) — any of a group of soft, simple marine animals with tentacles and a flowerlike appearance

species (SPEE sheez) — within a group of closely-related animals, such as jellyfish, one certain kind or type (*moon* jellyfish)

tentacles (TEN tah kulz) — a group of long (usually), flexible body parts generally growing around an animal's mouth and used for touching, grasping or stinging

transparent (tranz PARE ent) — able to be seen through; clear or nearly clear

INDEX